best of british
TRACTORS

best of british
TRACTORS

First published in the UK in 2015

www.g2books.co.uk

Printed and bound in Europe

ISBN 978-1-78281-928-8

Contents

Chapter 1

Allis-Chalmers

Allis-Chalmers was formed when Edward Allis, Fraser & Chalmers, Gates Iron Works, and Dickson's Manufacturing merged in 1901. Edward Allis was an entrepreneur who bought struggling businesses and turned them round. His company began producing steam engines in the 1860s but it went into liquidation in 1873. He bounced back and by the time his sons, Charles and William, took over in 1889 it was prosperous once more. By 1900, the Allis Company was the largest builder of steam engines in North America.

Thomas Chalmers was a Scottish immigrant who arrived in the US in around 1842. He soon found himself running a business building ploughs and farm machinery, mining equipment and boilers. The Dickson family, meanwhile, were forging a reputation as a machine company specialising in locomotives and internal combustion engines.

After the famous merger they enjoyed a decade of prosperity but financial difficulties led to a reorganisation in 1912. The new Allis-Chalmers Manufacturing Company entered the farm machinery business and introduced its first tractor, the 10-18, in 1914. The 1920s brought widespread mechanisation to farms across the continent and by the end of the decade tractors and farm equipment accounted for 60% of sales.

The company enjoyed continued success in the 1930s with its WC, B and C models. Allis-Chalmers were awarded defence contracts during World War Two, and the technological advances made during the conflict allowed them to provide machines with far more power throughout the 1950s and '60s. The dollar was weak against the pound at the time so a manufacturing base was established in Totton in the UK. Offices were then set up in Herefordshire to manage the European end of the business.

Farm equipment operations were sold to German brand Klockner-Humboldt-Deutz in 1985 but tractor production continued under the Deutz-Allis name until 1989. The company was then sold again and renamed AGCO. This company continued building tractors under the AGCO-Allis name until 2001.

Allis-Chalmers
MODEL B

The Allis-Chalmers Model B was designed by Brooks Stevens and built by Harry Merritt, manager of Allis-Chalmers' tractor division in the early 1930s. Most tractors were being used on farms of more than 100 acres but the Allis B was marketed to farmers with smaller holdings and it became one of the best-selling tractors in the company's history. With 125,000 units sold, it finally put the horse out of business on most small farms. The Model B was a light and versatile rowcrop tractor that required less maintenance than horses and was therefore cheaper. Such was its success that the company introduced an even smaller tractor, the Model G, in 1948.

SPECIFICATION

Fuel: **Petrol**

Engine capacity: **2.1 litres**

Cylinders: **4**

Power: **22hp**

Operating RPM: **400 - 1400**

Weight: **934kg**

Top speed: **8mph**

Length / width / height: **2.80 / 1.34 / 1.95** (m)

Years of production: **1938 - 1957**

Allis-Chalmers
D270

The D270 replaced the English version of the Model B in July 1954. There were few differences between the two, with only the tinwork (rolled metal on the body) being altered, as well as a few small upgrades to the linkage (a snap-coupler was added) and powertrain. Two engines were available: a petrol-paraffin version pumped out 23 horsepower while a later diesel variant boasted 27 horsepower from its Perkins P3-TA engine. As with the Model B, it had a four-speed transmission, 12-volt electric starter and a hydraulic system that was soon improved.

SPECIFICATION

Fuel: **Petrol**

Engine capacity: **2.1 litres**

Cylinders: **4**

Power: **23hp**

Operating RPM: **400 - 1650**

Weight: **1134kg**

Top speed: **12mph**

Length / width / height: **3.04 / 2.12 / 1.98** (m)

Years of production: **1954 - 1957**

Allis-Chalmers
D272

The D272 was marketed as an all-new tractor but it was really a 270 with more small upgrades. A three-cylinder Perkins diesel was available with 31 horsepower but the trend at the time was to use the four-pot petrol (28hp) or the petrol-paraffin mix (26hp) that was known as TVO (tractor vaporising oil). It was marketed as the 'tractor you couldn't afford to be without' as it promised to make better use of your existing tools, had heavy-duty independent brakes, a swinging drawbar, independent power takeoff and optional power-assisted variable tread (PAVT).

SPECIFICATION

Fuel: **Petrol**
Engine capacity: **2.1 litres**
Cylinders: **4**
Power: **28hp**
Operating RPM: **400 - 1650**
Weight: **934kg**
Top speed: **13mph**
Length / width / height: **2.79 / 1.98 / 1.60** (m)
Years of production: **1957 - 1960**

Allis-Chalmers
ED40

The ED40 was the all-new tractor the British farming community had been waiting for. It was unveiled to local dealers in Harrogate in late November 1960 and made its public debut at the Smithfield Show in December. It was powered by a 2.3-litre direct-injection diesel engine that had been developed by Harry Ricardo (which was similar to that used in the Ferguson 35) but it suffered in the cold and wasn't the company's most reliable unit.

It had a dual-range four-speed gearbox giving eight forward speeds and two in reverse. As automatic draft control was still covered by Ferguson's three-point patents, the ED40 used a rather crude adjustable stop to control the plough depth. At £622 for the top-of-the-range model with cushioned seat, swinging drawbar and tractorometer (a combined rev counter and hour meter), it wasn't the cheapest tractor either.

SPECIFICATION

Fuel: **Diesel**
Engine capacity: **2.3 litres**
Cylinders: **4**
Power: **41hp**
Operating RPM: **600 - 2250**
Weight: **964kg**
Top speed: **17mph**
Length / width / height: **2.89 / 1.82 / 1.90** (m)
Years of production: **1960 - 1968**

Chapter 2

Bristol

Bristol originally produced small crawler tractors in partnership with the Douglas Motorcycle Company, but they also worked alongside the Jowlett car manufacturer. The idea was to produce lightweight crawler tractors that would help local authorities care for verges, market gardeners tend to crops and ground staff to keep recreation areas in perfect condition. When Marshall bought the company, it became the Track Marshall Group.

Bristol
10

The Bristol 10 (on account of the horsepower the little Jowett flat-twin engine produced) wasn't a success but it paved the way for Bristol to make waves in the crawler market. They teamed with Roadless Traction in Hounslow to produce a pair of prototypes but the company didn't have the funds to develop them further or bring them to market. In 1937, William Jowett bought Bristol Tractors and launched the new and improved version for just £195. The company found its engines in great demand during the war so it took until 1947 for the next incarnation to be released, by which time the 10 was obsolete.

SPECIFICATION

Fuel: **Petrol**

Engine capacity: **1.2 litres**

Cylinders: **2**

Power: **10hp**

Operating RPM: **400 - 750**

Weight: **450kg**

Top speed: **5mph**

Years of production: **1933 - 1937**

Bristol 20

The Bristol 20 was far more successful in terms of usability and units sold. The Austin engine was bolted to the transmission case, there were separate levers for each steering clutch and independent foot pedals operating the brakes. The tracks were upgraded in-house at the factory in Earby near Bradford in Yorkshire following the dissolution of the agreement with Roadless. The linkage was also redesigned so that it could accommodate a two-furrow plough, and a guide wheel was added for accuracy and stability. As such, it was capable of more strenuous agricultural or industrial work.

SPECIFICATION

Fuel: **Petrol / TVO**

Engine capacity: **2.2 litres**

Cylinders: **4**

Power: **23hp**

Operating RPM: **1500**

Weight: **600kg**

Top speed: **8mph**

Years of production: **1949 - 1953**

Bristol 25

The Bristol 25 arrived in 1955 but it was designed on the cheap and only had three spaced bottom rollers, although the steel tracks had 10-inch pads as standard. The crawler came in an orchard/vineyard configuration – the tracks were enclosed with fenders to prevent snagging on the crops or damaging low branches – or as a civil engineering version with a loading bucket. The D version was launched in 1959 and, when the Perkins P3-144 Diesel engine arrived in 1960, the Model 25 was gradually replaced.

SPECIFICATION

Fuel: **Petrol**
Engine capacity: **2.2 litres**
Cylinders: **4**
Power: **32hp**
Operating RPM: **1500**
Weight: **630kg**
Top speed: **9mph**
Years of production: **1955 - 1959**

Chapter 3

David Brown

David Brown was a patternmaker who started a business manufacturing mechanised gears in Huddersfield in 1860. By 1898 the company was specialising in machine-cut gears so they moved to bigger premises at Park Works. David died in 1903 so his sons, Percy and Frank, continued to expand the business into bearings and shafts that would be used for the propulsion units in warships and submarines in the Great War. Business was predictably good during the conflict and by 1921 the company was the largest of its kind in the world.

Percy's son, Sir David Brown, took over in 1931 and he joined forces with Harry Ferguson to build Ferguson-Brown tractors from 1936. The partnership soon dissolved, however, and Ferguson left for the United States to team up with Henry Ford and the Ford Motor Company.

The first David Brown tractor, the VAK1, was released at the Royal Show in 1939. More than 3,000 units were ordered in the UK but the outbreak of the Second World War meant that only around a third of the orders were fulfilled. Production resumed after the war and the business was eventually taken over by Tenneco and operated as a division of J.I. Case from 1972. By 1983, tractors were no longer being built under the David Brown name, and the factory in Meltham Mills where all their machinery had been built was closed five years later.

David Brown
CROPMASTER VAK1/C

Having split from Harry Ferguson, David Brown introduced the VAK1 in their now-familiar 'hunting pink' livery. The tractor finally took on the shape and configuration on which all modern tractors are based: four wheels (no tyres yet though), four cylinders in a standard internal combustion engine, and a four-speed gearbox. The Second World War forced David Brown to explore other manufacturing options, which consisted largely of defence contracts, but they still managed to produce 5,000 units and their tractors evolved quickly after the conflict as a result of all the research and development.

By 1947, the company had released its seminal machine: the VAK1/C Cropmaster. In only six years, this iconic tractor would shift 60,000 units worldwide, and it would also spawn countless derivatives and imitations.

SPECIFICATION

Fuel: **Kerosene**
Engine capacity: **2.5 litres**
Cylinders: **4**
Power: **25hp**
Operating RPM: **600 - 2000**
Weight: **1338kg**
Top speed: **15mph**
Length / width / height: **2.83 / 1.53 / 1.40** (m)
Years of production: **1947 - 1953**

David Brown CROPMASTER

The long production run of David Brown's Cropmasters gave the company the opportunity to pioneer new features so their tractors earned a reputation for reliability and quality. With the introduction of the Cropmaster diesel, the company decided to supply the standard machine with all the items normally bought as extras, such as the hydraulic lift and swinging drawbar.

A two-speed power takeoff, six-speed gearbox, coil ignition and high-speed direct-injection diesel engine were also introduced, and power was increased from 30 to 34 horsepower in 1952 when the engine was re-bored. This diesel engine was lighter and more compact than previous incarnations and it proved to be more reliable than the earlier petrol or kerosene powerplants. As it saved on fuel, oil and other maintenance costs, the Cropmaster diesel was also more economical to own.

SPECIFICATION

Fuel: **Diesel**
Engine capacity: **2.5 litres**
Cylinders: **4**
Power: **34hp**
Operating RPM: **700 - 1800**
Weight: **1338kg**
Top speed: **16mph**
Length / width / height: **2.83 / 1.53 / 1.40** (m)
Years of production: **1949 - 1953**

David Brown 50D

In April 1953 the Cropmaster's successful run was brought to an end by the latest variant, the 50D. Although it was very similar to the model it replaced – six forward speeds, the economy and reliability of a diesel, side-mounted pulley, provision for 4-6 furrows and reserve power for driving combines, balers and other heavy machinery – the 50D also had a side-mounted oil-bath and air cleaner. As it didn't have a differential lock, hydraulic lift or linkage it was primarily exported overseas for haulage work and only 1,260 were sold in the UK before production ended in 1958.

SPECIFICATION

Fuel: **Diesel**
Engine capacity: **4.1 litres**
Cylinders: **6**
Power: **50hp**
Operating RPM: **700 - 1800**
Weight: **2780kg**
Top speed: **13mph**
Length / width / height: **3.42 / 1.87 / 1.65** (m)
Years of production: **1953 - 1958**

David Brown
25D

The 25 series was designed to be a more economical and manoeuvrable tractor than the Cropmasters it replaced, so it was stripped of its second seat, horseshoe scuttle and bumpers. Although it was sold as a basic farm machine, it had a six-speed gearbox, two-speed power takeoff and a hydraulic lift. The clutch pedal was shifted to the left and the driver sat in the centre above the transmission (unlike previous models that had an offset driving position). The 25D (diesel) was so popular – and cheap at only £470 – that more than 13,000 units were sold in five years.

SPECIFICATION

Fuel: **Diesel**

Engine capacity: **2.5 litres**

Cylinders: **4**

Power: **32hp**

Operating RPM: **600 - 1800**

Weight: **1685kg**

Top speed: **14mph**

Length / width / height: **2.87 / 1.72 / 1.44** (m)

Years of production: **1953 - 1958**

David Brown
30C

The petrol-powered 30C was another replacement for the outdated Cropmaster, although some of the early machines used the same tinwork (rolled metal structure). From 1954 the 30C reverted to a more similar styling as the 25 series but a new traction-control unit was then installed. It had the same 2.7-litre engine as the older Super Cropmaster but the company would eventually follow convention and run with diesel engines as they saved on fuel, were more reliable and had lower overall running costs.

SPECIFICATION

Fuel: **Petrol**
Engine capacity: **2.7 litres**
Cylinders: **4**
Power: **41hp**
Operating RPM: **2300**
Weight: **1769kg**
Top speed: **15mph**
Years of production: **1953 - 1957**

David Brown
2D

The 2D rowcrop tractor was innovative in every sense: it had a compressed-air lift, two-cylinder air-cooled diesel engine with another idler cylinder to prevent vibration, and a four-speed gearbox. As it was a specialist rowcrop machine and weighed much less than a tonne, it appealed to market gardeners and sold 2,000 units in its five-year production run.

SPECIFICATION

Fuel: **Diesel**
Engine capacity: **1.3 litres**
Cylinders: **2**
Power: **14hp**
Operating RPM: **600 - 1800**
Weight: **861kg**
Top speed: **9mph**
Length / width / height: **2.95 / 1.70 / 1.39** (m)
Years of production: **1956 - 1961**

David Brown
900

The 900 was launched towards the end of 1956 and it was the first British tractor to boast a CAV rotary pump. Although it was supposed to drag the company into a new and exciting era, the injection pump hadn't been properly tested and it repeatedly failed, causing David Brown to recall faulty models. It wasn't a complete disaster, however, and in 1957 the 900 was upgraded to include a 'livedrive' dual-clutch system that allowed the driver to operate the power takeoff and hydraulics independently of the tractor's speed.

This version was the first to break with the company's traditional pink livery in that the model had blue wheels. As well as the diesel version, orders could be placed for petrol and tractor vaporising oil, although the diesel was by far the most popular.

SPECIFICATION

Fuel: **Diesel**
Engine capacity: **2.7 litres**
Cylinders: **4**
Power: **42hp**
Operating RPM: **2200**
Weight: **1814kg**
Top speed: **15mph**
Length / width / height: **2.87 / 1.65 / 1.97** (m)
Years of production: **1956 - 1958**

David Brown
950

Having had so many problems with the 900, the 950 was a breath of fresh air. Launched in November 1958, it also came with livedrive and a petrol option (mainly for the export market), but its 43-horsepower direct-injection diesel was again the most popular variant. It had lighter steering, a higher ground clearance, universal linkage with rust-resistant pins, optional towing drawbar, and 2,000lb hydraulics for lifting and loading.

SPECIFICATION

Fuel: **Diesel**

Engine capacity: **2.7 litres**

Cylinders: **4**

Power: **43hp**

Operating RPM: **700 - 2200**

Weight: **1918kg**

Top speed: **12mph**

Length / width / height: **3.00 / 1.63 / 1.42** (m)

Years of production: **1958 - 1959**

David Brown
850 IMPLEMATIC

Developed by hydraulics experts Harry Horsfall and Charles Hull, the 850 Implematic drew on the best of the four-cylinder engines from the previous 25/25D series and the features and specifications from the 950. As such it was a perfect hybrid of the two, with a six-speed gearbox, power steering, hydraulic isolator valve, combined tacho and hourmeter, and a selective draft control sensor for easier ploughing. The livedrive model was £30 more expensive but it was something of a bargain at £644.

SPECIFICATION

Fuel: **Diesel**

Engine capacity: **2.5 litres**

Cylinders: **4**

Power: **35hp**

Operating RPM: **2000**

Weight: **1895kg**

Top speed: **15mph**

Length / width / height: **2.87 / 1.59 / 1.31** (m)

Years of production: **1960 - 1965**

David Brown
880 IMPLEMATIC

The David Brown 880 was something of a stop-gap as a new three-cylinder engine was in the pipeline. There were three production versions of the 880 Implematic but the first only lasted until the new engine was launched in 1964. It had previously been used in the 30D and was now rated at 43 horsepower. Both the 850 and 880 Implematic remained in production alongside one another until the paint scheme and hydraulics were upgraded in 1965 to produce the 880 Selectamatic.

SPECIFICATION

Fuel: **Diesel**

Engine capacity: **2.7 litres**

Cylinders: **4**

Power: **40hp**

Operating RPM: **1800**

Weight: **1957kg**

Top speed: **15mph**

Length / width / height **2.92 / 1.62 / 1.33** (m)

Years of production: **1961 - 1965**

David Brown
880 SELECTAMATIC

The 880 was fitted with a selectamatic hydraulic system and was delivered in the restyled orchid white livery. The powerplant now developed 46hp and the 880 came with a multi-speed PTO and differential lock as standard. It had 12 forward and four reverse gears for precision fieldwork and it also came with non-standard equipment like optional four- or six-ply tyres. It could be converted from low to high axle clearance for undertaking tougher work on uneven ground. In 1970, a full-flow filtration system for the hydraulic oil was introduced.

SPECIFICATION

Fuel: **Diesel**

Engine capacity: **2.7 litres**

Cylinders: **3**

Power: **46hp**

Operating RPM: **700 - 2200**

Weight: **2027kg**

Top speed: **15mph**

Length / width / height: **3.09 / 1.66 / 2.33** (m)

Years of production: **1961 - 1971**

David Brown
990 IMPLEMATIC

The 990 Implematic went into production in September 1961 with a four-cylinder diesel engine. New features included a cross-flow head with inlet ports for each cylinder on one side and exhaust outlets on the other side of the head. This helped ensure clean combustion and reduced the amount of hot exhaust gases that mixed with the air-induction process.

From 1964, the battery was repositioned in front of the radiator to improve lighting. A larger toolbox came as standard and an optional 12-speed gearbox was introduced. The oil bath and air cleaner were also mounted inside the radiator grille.

SPECIFICATION

Fuel: **Diesel**
Engine capacity: **3 litres**
Cylinders: **4**
Power: **52hp**
Operating RPM: **700 - 2200**
Weight: **1937kg**
Top speed: **12mph**
Length / width / height: **2.99 / 1.70 / 1.42** (m)
Years of production: **1961 - 1965**

David Brown
770 SELECTAMATIC

The 770 Selectamatic was launched at the Smithfield Show in late 1964 but it didn't go into production until the following year. The three-cylinder AD3/40 engine that had been used in the 880 was given a shorter stroke but the overall styling changed little from the existing models. The 770 was the last model to be delivered with the company's iconic 'hunting pink' paint job and yellow wheels.

The controls and valves for the tractor's hydraulics were positioned in a single case with one switch for easier use. This system would be incorporated on all new David Brown tractors from 1965, which were then given a distinctive white livery to distinguish them from earlier models.

SPECIFICATION

Fuel: **Diesel**
Engine capacity: **2.4 litres**
Cylinders: **3**
Power: **28hp**
Operating RPM: **600 - 1800**
Weight: **1723kg**
Top speed: **11mph**
Length / width / height: **3.20 / 1.65 / 2.03** (m)
Years of production: **1965 - 1970**

David Brown
780

The DB 780 was essentially a 770 with the engine from an 880. It was introduced in 1967 because of demand from dealers and farmers wanting the performance of the 880 with the specifications of the 770. The pink livery had been swapped for white but customers were initially lukewarm about the new tractor and it only sold 878 units in its first year. However, the selectamatic gearbox option plus the introduction of a narrow version and the addition of a cab and safety frame saw demand increase rapidly. By the end of 1968, David Brown had shifted 2,700 units, and another 3,300 were sold in each of the next two years.

SPECIFICATION

Fuel: **Diesel**
Engine capacity: **2.7 litres**
Cylinders: **3**
Power: **46hp**
Operating RPM: **600 - 1400**
Weight: **1573kg**
Top speed: **13mph**
Length / width / height: **2.83 / 1.62 / 2.09** (m)
Years of production: **1967 - 1971**

David Brown
3800 SELECTAMATIC—

DB tractors continued to evolve with the 3800, which was essentially a 780 with a 2.4-litre 39-horsepower petrol engine rather than the 2.7-litre 45-horsepower diesel. Many versions came with a front loader and three-point hitch-post hole digger. The PTO was often driven from an aftermarket hydraulic pump at the front of the crankshaft, although DB wouldn't have set up a factory unit in such a way as owners usually modified them. Only around 500 were ever made so very few remain in operation today.

SPECIFICATION

Fuel: **Petrol**
Engine capacity:
2.4 litres
Cylinders: **3**
Power: **39hp**
Operating RPM:
700 - 2200
Weight: **1762kg**
Top speed: **17mph**
Length / width / height:
2.93 / 1.63 / 2.09 (m)
Years of production:
1968 - 1971

David Brown
4600 SELECTAMATIC

This utility tractor used a three-cylinder liquid-cooled petrol engine so it was primarily for the lucrative export market, particularly North America. With only 140 foot-pounds (lb-ft) of torque, it wasn't powerful enough for hard fieldwork, however, and petrol engines were less economical than diesel. For these reasons, only 582 examples were sold (serial numbers 900001 - 900582).

SPECIFICATION

Fuel: **Petrol**
Engine capacity: **2.7 litres**
Cylinders: **3**
Power: **46hp**
Operating RPM: **700 - 2200**
Weight: **1950kg**
Top speed: **16mph**
Length / width / height **3.08 / 1.73 / 1.37** (m)
Years of production: **1968 - 1971**

David Brown
885

The 12 forward and four reverse gears (with synchromesh) became standard on all David Brown tractors (except the 1212) in 1971. The 885 incorporated the best features from both the 780 and 880, and it was also available as a narrow version. Several manufacturers adopted DIN engine ratings in 1972 so the 885 became 48 DIN instead of 47hp, although for consistency across all manufacturers horsepower has been used throughout this book. A safety weather-frame (usually the Sekura lo-profile cab) was introduced in 1971 and an alternator was included with the standard unit in 1973. With its petrol engine appealing to buyers overseas, the 885 did well on the export market.

SPECIFICATION

Fuel: **Petrol**
Engine capacity: **2.4 litres**
Cylinders: **3**
Power: **47hp**
Operating RPM: **2200**
Weight: **1934kg**
Top speed: **16mph**
Length / width / height: **2.94 / 1.65 / 2.38** (m)
Years of production: **1971 - 1980**

David Brown
990 SELECTAMATIC

Most 990 tractors were manually driven but some were fitted with the optional hydrostatic cylinder to provide power assistance. The orbital valve in the hydrostatic steering models controlled the steering cylinder. David Brown stuck with the popular 12 forward and four reverse gears configuration, while front axle clearance was raised to 54cm. The hydraulic pressure in the system was kept constant at 2,000psi but the maximum static lift was increased to 1,930kg.

SPECIFICATION

Fuel: **Diesel**
Engine capacity: **3.2 litres**
Cylinders: **4**
Power: **55hp**
Operating RPM: **700 - 2200**
Weight: **2313kg** (2WD) **2628kg** (4WD)
Top speed: **14mph**
Length / width / height: **3.22 / 1.62 / 2.41** (m)
Years of production: **1965 - 1980**

David Brown
995 & 996

The DB 995 and 996 had slightly larger engines than the 990 but many of the features remained the same: the universal drawbar could be laterally adjusted up to 150mm from the centre; it had three forward positions; and its maximum load was 1,800kg. An uprated safety cab was introduced to reduce noise to 90 decibels and a battery with 96 amp hours was included as standard. Accessories included rear wheel weights, a pick-up hitch, linkage stabilisers, support ram, radiator guard, tool kit, radio and luxury seat covers. Ground clearance for the front axle could be raised to 730mm in high-clearance configuration.

SPECIFICATION

Fuel: **Diesel**
Engine capacity: **3.6 litres**
Cylinders: **4**
Power: **64hp**
Operating RPM: **700 - 2200**
Weight: **2612kg**
Top speed: **12mph**
Length / width / height: **3.30 / 1.70 / 2.51** (m)
Years of production: **1971 - 1980**

A David Brown 996

David Brown
1200 SELECTAMATIC

Farmers were always looking for more grunt so David Brown launched the 1200 Selectamatic in 1967 with a 72 horsepower four-cylinder direct-injection diesel engine coupled to six forward and two reverse gears (a 12-speed box was available as an option). Although this raised the top speed by two miles per hour, it was the increase in torque that mattered when drawing loads. Standard equipment included independent power takeoff with a hand-operated clutch, selectamatic hydraulics, drum brakes and a pedal-operated exhaust brake (the hydraulic pump was mounted in front of the engine with a three-point linkage). This pneumatic engine brake restricted the flow of exhaust gases from the engine and retarded power to the rear wheels. A luxury seat was fitted as standard on the four-wheel-drive version that was launched in 1970.

SPECIFICATION

Fuel: **Diesel**
Engine capacity:
3.6 litres
Cylinders: **4**
Power: **72hp**
Operating RPM:
700 - 2300
Weight: **2825kg** (2WD)
3034kg (4WD)
Top speed: **14mph**
Length / width / height:
3.63 / 1.87 / 2.36 (m)
Years of production:
1967 - 1971

David Brown
1210

The DB 1210 employed standard synchromesh transmission with 12 forward and four reverse speeds. It came as both 2WD and 4WD, although the latter was 10% slower in all gears. In the 4WD version, a hand lever allowed the driver to switch back to 2WD to increase economy on roads, and the single central driveshaft ran directly from the gearbox to the front axle differential. Hydrostatic power steering came as standard on the 4WD but was a factory option on the 2WD.

SPECIFICATION

Fuel: **Diesel**
Engine capacity: **3.6 litres**
Cylinders: **4**
Power: **56hp**
Operating RPM: **700 - 2200**
Weight: **2960kg** (2WD) **3190kg** (4WD)
Top speed: **13mph**
Length / width / height: **3.65 / 1.87 / 2.60** (m)
Years of production: **1971 - 1976**

David Brown
1212

The DB 1210 and 1212 models were surprisingly similar in configuration and performance but the main difference was in the transmission: the 1212 employed semi-automatic hydra-shift technology delivering clutchless changes between three forward and one reverse gear while on the move. Engine braking was available in every gear, while the tractor could also be tow-started or even towed with a dead engine. It had hydrostatic power steering and nearly 20cm greater clearance at the front axle than the 4WD 1210.

SPECIFICATION

Fuel: **Diesel**
Engine capacity: **3.6 litres**
Cylinders: **4**
Power: **71hp**
Operating RPM: **700 - 2300**
Weight: **3050kg**
Top speed: **16mph**
Length / width / height: **3.65 / 1.77 / 2.55** (m)
Years of production: **1971 - 1976**

David Brown

1410

The DB 1410 used a direct-injection four-cylinder diesel engine with a compression ratio of 16:1. It also had the now-standard cross-flow cylinder head, overhead valves and a balanced crankshaft, as well as an inline fuel pump, dry-air cleaner, service indicator and cold-start aid. The semi-automatic hydra-shift transmission was also retained, and the disc brakes could be operated separately or together via a master pedal. Rollovers were rare but David Brown continually improved safety with a weather cab that included safety-glass doors, front panels, a folding rear curtain and heavy-duty wipers.

SPECIFICATION

Fuel: **Diesel**

Engine capacity: **3.6 litres (turbocharged)**

Cylinders: **4**

Power: **91hp**

Operating RPM: **700 - 2300**

Weight: **3474kg**

Top speed: **16mph**

Length / width / height: **3.68 / 1.95 / 2.61** (m)

Years of production: **1974 - 1979**

David Brown

1412

The first production 1412s had turbocharged 3.6-litre direct-injection diesel engines rated at 91 horsepower. Other new features included a dry-air cleaner, oil-immersed disc brakes, telescopic lower-link ends, and a balanced hydrostatic steering ram that was mounted behind the front axle. The 1412 was equipped with the 12-speed semi-automatic hydra-shift gearbox as standard. The 1410s and 1412s divided opinion, with some farmers claiming they – along with the 900s – were the worst ever produced by David Brown. They cite burning the tops of the pistons under hard work, cracked blocks, faulty head gaskets and replacing the clutches and hydra-shift systems as expensive weaknesses. Many other farmers, however, have had few problems even after 12,000 hours of use.

SPECIFICATION

Fuel: **Diesel**

Engine capacity: **3.6 litres (turbocharged)**

Cylinders: **4**

Power: **91hp**

Operating RPM: **700 - 2300**

Weight: **3474kg**

Top speed: **16mph**

Length / width / height: **3.68 / 1.95 / 2.61** (m)

Years of production: **1974 - 1979**

David Brown / J.I. Case

1390

The DB 1390 was introduced at the same time as the similar but less powerful 1290. It had 2WD and 4WD options and a reliable direct-injection four-pot diesel engine. Standard equipment included a luxury safety cab, 12-speed synchromesh gearbox in four overlapping ranges (with another four reverse gears), hydrostatic steering, single-lever hydraulics, multi-speed PTO, a hydraulically operated clutch, and self-balancing brakes. Some tractors were now using oiled disc brakes but the 1390 retained hydraulic drum brakes with separate pedals that could be latched together for road use.

SPECIFICATION

Fuel: **Diesel**

Engine capacity: **3.6 litres**

Cylinders: **4**

Power: **60hp**

Operating RPM: **700 - 2200**

Weight: **2494kg**

Top speed: **15mph**

Length / width / height: **3.42 / 1.87 / 2.23** (m)

Years of production: **1980 - 1984**

David Brown / J.I. Case
1490

The 1490 used essentially the same configuration and engine as the 1390 but David Brown added a turbocharger to increase power by 23hp. The PTO clutch had a separate lever while the foot-operated clutch was hydraulically powered. The 1490 also benefitted from the semi-automatic 12-speed hydra-shift transmission, while a simple lever on the 4WD variant allowed the driver to switch to 2WD for lighter work or road use. More controls like a starter isolating switch, temperature gauge and horn were also added, as were more effective disc brakes with twin foot pedals.

SPECIFICATION

Fuel: **Diesel**

Engine capacity: **3.6 litres (turbocharged)**

Cylinders: **4**

Power: **83hp**

Operating RPM: **700 - 2200**

Weight: **2948kg** (2WD) **3202kg** (4WD)

Top speed: **15mph**

Length / width / height: **3.63 / 1.94 / 2.66** (m)

Years of production: **1980 - 1984**

David Brown / J.I. Case

1690

While the 1490 was turbocharged, David Brown went with a naturally aspirated 5.4-litre six-cylinder diesel for the 1690, which was sold over the same five-year period. The larger engine allowed many users to modify the powertrain and some managed to increase power to more than 135hp. Many of the tried-and-tested features remained the same as on the 1490, notably the 12 forward and four reverse gears, the clutchless hydra-shift box, a reversible independent PTO shaft with 540 and 1000 splines and a standard front guard.

SPECIFICATION

Fuel: **Diesel**
Engine capacity:
5.4 litres
Cylinders: **6**
Power: **103hp**
Operating RPM:
700 - 2200
Weight: **3855kg** (2WD)
4093kg (4WD)
Top speed: **15mph**
Length / width / height:
3.95 / 2.00 / 2.60 (m)
Years of production:
1980 - 1984

A David Brown Cropmaster

Rear view of a cutaway display of a 1953 David Brown 30T Trackmaster track-laying tractor

Straw baling using a David Brown 1410 and Jones baler

A David Brown / J.I. Case 1490

Chapter 4

Ferguson

Harry Ferguson was born in County Down, Ireland, in 1884. A keen inventor and engineer, he experimented with various types of tractor and plough in the early years of the 19th century. He then started a tractor business selling Waterloo Boys during the First World War. He continued working with farmers after the conflict and helped design a better system to attach all the implements to farm and utility tractors, which would become known as the three-point hitch. To demonstrate the new system, Ferguson designed the Black tractor in the 1930s. This came to the attention of renowned manufacturer David Brown and the pair joined forces to build the Ferguson-Brown.

The arrangement didn't last however, and, as Brown was already working on the VAK1, Ferguson headed for America and shook on a deal with Henry Ford that would allow him to continue developing tractors using his hitch. All Ford tractors produced from the 9N (1939-1942) and the 2N (1942-1947) used Ferguson's system and he received a share of the profits as part of a licence deal.

In 1947, the deal with Ford ended in an acrimonious lawsuit when Henry's grandson terminated the agreement, so Ferguson withdrew his support for the latest 8N range and teamed up with the Standard Motor Company to produce their tractors instead. (Ferguson eventually won a substantial payout – $9.25 million – from Ford for the way the agreement was ended.) Ferguson also hit back by producing rival machines, beginning with the TO-20 in 1948. Five years later, his company was taken over by Massey-Harris but the Massey-Ferguson brand is still in the public consciousness today as part of the global conglomerate AGCO.

Ferguson was the first Irishman to design and fly his own aeroplane, and he also developed the first four-wheel-drive Formula One car, the P99. He died aged 75 in Gloucestershire, England.

Ferguson TYPE A

Following the famous Ferguson Black tractor, the Ferguson-Brown Type A was the first production tractor with the hydraulic three-point linkage hitch that would go on to become one of the most important innovations in mechanised farming. The tractor also needed a suction side hydraulic valve to control the hitch, so Ferguson approached gear-maker David Brown, and it was he who helped develop the system.

It made ploughing easier and more consistent, so the hitch was soon being adapted for every model of tractor in the world. The Type A may have only been a simple 20-horsepower petrol machine that sold around 1,300 units but it revolutionised farming across the globe. It was discontinued in the run-up to the Second World War.

SPECIFICATION

Fuel: **Petrol**

Engine capacity: **2 litres**

Cylinders: **4**

Power: **20hp**

Operating RPM: **1400**

Weight: **760kg**

Top speed: **9mph**

Years of production: **1936 - 1937**

Ferguson
TE-20

The TE-20 (tractor, England, 20 horsepower) was Ferguson's most successful design, and it was the tractor that finally ousted the horse on smaller holdings. It may have been small and underpowered but it had the three-point hitch and suction control valve that made ploughing so much more economical. Ferguson owned the patent so he entered a gentleman's agreement with Henry Ford to build Ford Fergusons using his production lines in Detroit. Up to 300,000 tractors were produced before the Ferguson-Ford alliance broke up in 1947. This left Ferguson without any tractors to sell so he imported TE-20s under the designation TO (tractor overseas). Most versions used petrol engines as they were more popular in North America, Europe and the Commonwealth, although some were supplied with tractor vaporising oil (a petrol-paraffin or kerosene mix) or a Perkins three-cylinder diesel engine.

SPECIFICATION

Fuel: **Petrol**
Engine capacity:
2 litres
Cylinders: **4**
Power: **21hp**
Operating RPM:
700 - 1750
Weight: **1251kg**
Top speed: **12mph**
Length / width / height:
2.92 / 1.61 / 1.32 (m)
Years of production:
1946 - 1948

Ferguson TE-A20

The Standard Motor Company's original 80mm bore petrol engine was used in TE-A20s until 1951. The tractor also came in a narrow or vineyard version, which was small enough to fit between rows of vines, so it was extremely popular on the continent. It was such a versatile tractor that explorers Dr Vivian Fuchs and Sir Edmund Hillary (fresh from his first ascent of Mount Everest) chose three converted TE-A20s for their historic overland crossing of Antarctica between 1955 and 1958. When Hillary arrived at the South Pole in January 1958, his became only the third expedition (after Amundsen and Scott) to reach 90 degrees south, and it marked the first time vehicles had ever reached the pole.

SPECIFICATION

Fuel: **Petrol**
Engine capacity: **1.9 litres**
Cylinders: **4**
Power: **24hp**
Operating RPM: **400 - 2000**
Weight: **1134kg**
Top speed: **13mph**
Length / width / height: **2.92 / 1.62 / 1.32** (m)
Years of production: **1948 - 1956**

Ferguson TE-F20

The TE-F20 was the diesel version of the A20. (A D20 had also been introduced in 1949, although this was less popular as it ran on tractor vaporising oil.) Both the A and F20 were picked up by building trades, local authorities and municipal contractors as it was extremely well suited for refuse collection, mowing grass verges, maintaining recreation grounds and transporting heavy loads across boroughs.

SPECIFICATION

Fuel: **Diesel**
Engine capacity: **2.1 litres**
Cylinders: **4**
Power: **26hp**
Operating RPM: **500 - 2000**
Weight: **1225kg**
Top speed: **13mph**
Length / width / height: **2.92 / 1.62 / 1.32** (m)
Years of production: **1951 – 1956**

Ferguson FE-35

The Ferguson FE-35 went into production as the Ferguson and Massey-Harris companies merged into Massey-Harris-Ferguson (the Harris was dropped between 1958 and 1960). The tractor began life in grey-and-gold paint with a Ferguson badge but the livery changed to red and grey in late 1957. It inherited much of the TE-20's DNA but also included a six-speed gearbox, a two-speed power takeoff and uprated hydraulics. The Standard Motor Company again provided a series of petrol, diesel and TVO engines, but Ferguson was struggling to shift units while marketing two models. The TE-20s were thus phased out while the FE-35 was re-launched in December 1957.

SPECIFICATION

Fuel: **Petrol**
Engine capacity: **2.2 litres**
Cylinders: **4**
Power: **34hp**
Operating RPM: **600 - 1800**
Weight: **1402kg**
Top speed: **14mph**
Length / width / height: **3.05 / 1.63 / 1.82** (m)
Years of production: **1956 - 1958**

Chapter 5

Ivel

English inventor Dan Albone was a gifted sprint cyclist who won nearly 200 major races on a bike he built himself. By 1880, the 20-year-old had founded the Ivel Cycle Works and was producing racing bikes and tricycles for local enthusiasts. In the 1890s he graduated to building motorcars and had exhibited his prototypes in Bedford and London. He also realised he could adapt his factory to produce farm machinery and built a three-wheel petrol light farm tractor in 1903. He died three years later but the model remained in production until 1920 when Ivel Agricultural Motors was dissolved.

Ivel Agricultural Motors

IVEL

With farms still being powered by horses and heavy, slow and inefficient traction engines, Dan Albone realised the older steam engines could be replaced with lighter internal combustion-driven tractors. He came up with a basic design in 1901 and filed for a patent the following year. In December 1902 he founded Ivel Agricultural Motors Limited and asked famous names from motor-racing such as Selwyn Francis Edge and Charles Jarrott to join the board of directors.

Albone's tractor won a ploughing competition in 1904 and 500 units were eventually built for the UK and export markets. Despite the innovative methods used in its construction, the company failed to keep up with a clutch of rivals and the Ivel had all but vanished by 1920.

SPECIFICATION

Fuel: **Petrol**

Engine capacity: **2.9 litres**

Cylinders: **2**

Power: **24hp**

Operating RPM: **750**

Weight: **1814kg**

Top speed: **5mph**

Years of production: **1903 - 1920**

Chapter 6

JCB

When Joseph Cyril Bamford rented a small garage in Staffordshire in 1945 and manufactured a tipping trailer from surplus war materials, he couldn't have envisioned how his company would grow over the next 70 years. By 1948 he was employing six people to build small hydraulic trailers but it wasn't long before they were making backhoes, excavators, diggers and tractors.

In 1960, the company delivered its first machines to North America and the brand name soon spread to India, Japan and Brazil. JCB now has 18 factories on four continents and employs a workforce of 11,000. It is the world's third-largest equipment manufacturer for the construction, demolition and agricultural industries and produces more than 300 models of digger, backhoe, tractor and diesel engine. The company is worth approximately £3 billion.

JCB
FASTRAC 135

The JCB Fastrac series of agricultural tractors were designed to work at higher speeds than conventional tractors. The early models all used a turbocharged six-litre Perkins Diesel engine and employed a transmission with 18 forward gears over three ranges and six in reverse. All Fastrac models had selectable four-wheel-drive, with drive to the front axle controlled by the driver through a hydraulic clutch. Two-speed power takeoff was fitted as standard to the rear and was an option on the front. Hydropneumatic suspension was also standard on the rear.

SPECIFICATION

Fuel: **Diesel**
Engine capacity:
6 litres (turbocharged)
Cylinders: **6**
Power: **147hp**
Operating RPM:
1400 - 2400
Weight: **6500kg**
Top speed: **53mph**
Length / width / height:
5.76 / 2.41 / 2.98 (m)
Years of production:
1993 - 1997

JCB
FASTRAC 155

The Fastrac 155 was extremely well equipped and came with all-wheel-drive, a pneumatic system, front linkage and forward PTO, front-axle suspension system, air-conditioned cab, hydraulic steering assist and extended-life coolant. The tractor also delivered nearly 450lb-ft of torque, almost three times the pulling capacity of the David Brown and Ferguson tractors from the 1970s. The 36 forward gears and fuel-tank capacity of 74 US gallons (280 litres) meant that it was economical to run and maintain.

SPECIFICATION

Fuel: **Diesel**

Engine capacity:

6 litres (turbocharged)

Cylinders: **6**

Power: **168hp**

Operating RPM:

1400 - 2400

Weight: **6501kg**

Top speed: **53mph**

Length / width / height:

5.76 / 2.41 / 2.99 (m)

Years of production:

1993 - 1997

JCB
FASTRAC 185

The Fastrac 185 arrived the year after the 155 and although the new engine was slightly smaller than the old Perkins six-litre unit, power was increased from 168 to 188 horsepower. This gave an extra 20% torque, up from 450 to 553lb-ft at 1400rpm. The gearing and top speed remained the same as the 155 and the hydraulics gave a rear-lift capacity of just over seven tonnes. Hydrostatic steering, mechanical front-wheel-drive and 4WD controls were also available, as were disc brakes.

SPECIFICATION

Fuel: **Diesel**

Engine capacity: **5.9 litres (intercooled turbo)**

Cylinders: **6**

Power: **188hp**

Operating RPM: **1400 - 2400**

Weight: **6501kg**

Top speed: **53mph**

Length / width / height: **5.76 / 2.41 / 2.99** (m)

Years of production: **1994 - 1997**

JCB
FASTRAC 1115

The JCB 1115 used the standard six-cylinder diesel but this was a smaller machine so it only produced 117 horsepower and the fuel tank was also smaller at 220 litres (58 US gallons). The 36 forward and 12 reverse gears were retained, as were the 4WD systems, but the rear-lift capacity was reduced to 6300kg. An air-conditioned cab and dry disc brakes provided comfort and safety. This was the first smaller Fastrac to use hydrostatic steering but it retained the two-speed power takeoff (at 540 and 1000rpm).

SPECIFICATION

Fuel: **Diesel**

Engine capacity: **6 litres (turbocharged)**

Cylinders: **6**

Power: **117hp**

Operating RPM: **1400 - 2400**

Weight: **5300kg**

Top speed: **37mph**

Length / width / height: **5.16 / 2.33 / 2.80** (m)

Years of production: **1995 - 1996**

JCB
FASTRAC 1115S

The 1115S had an extra 15 horsepower over its predecessor, even though the engine was the standard six-litre turbocharged unit manufactured by Perkins (1000-6 HR1). As with the 1115, the S had a standard locking differential for the rear axle and a Detroit No-Spin differential lock for the front axle. This helped the series maintain traction on soft, muddy or uneven surfaces, and was essential for avoiding getting stuck in rocky or boggy terrain.

SPECIFICATION

Fuel: **Diesel**
Engine capacity: **6 litres (turbocharged)**
Cylinders: **6**
Power: **132hp**
Operating RPM: **1400 - 2300**
Weight: **5501kg**
Top speed: **37mph**
Length / width / height: **5.16 / 2.33 / 2.80** (m)
Years of production: **1996 - 1998**

JCB FASTRAC 1135

The 1135 was released in 1995 with a new 1000-6 HR5 six-litre turbo-diesel. Torque in the 1115S had dropped to just 386lb-ft but the 18-horsepower hike in the 1135 saw torque also increase to 450lb-ft. Along with the precise gearing and hydraulics, torque is perhaps the most important figure farmers look for. The pulling/towing capacity of tractors is certainly helped with 4WD but it's the torque that dictates the amount of work the tractor can accomplish. The hydraulic system, PTO, steering, brakes and transmission were all up to the job so were retained on all the 1000 series models.

SPECIFICATION

Fuel: **Diesel**
Engine capacity: **6 litres (turbocharged)**
Cylinders: **6**
Power: **150hp**
Operating RPM: **1400 - 2300**
Weight: **5501kg**
Top speed: **37mph**
Length / width / height: **5.16 / 2.33 / 2.80** (m)
Years of production: **1995 - 1998**

JCB
FASTRAC 1125

The Fastrac 1125 used a slightly different 1000-6 HR2 engine that developed 141 horsepower from its turbocharged six-cylinder diesel. Torque was down on some of the earlier models to 430lb-ft at 1400rpm but the gearing (36 forward and 12 reverse) and top speed remained the same as with the rest of the 1000 series. PTO speeds also remained the same at 540/1000rpm, and the rear-lift capacity was still 6301kg. The 4WD system, hydrostatic steering and disc brakes were retained as standard.

SPECIFICATION

Fuel: **Diesel**
Engine capacity: **6 litres (turbocharged)**
Cylinders: **6**
Power: **141hp**
Operating RPM: **1400 - 2300**
Weight: **5501kg**
Top speed: **37mph**
Length / width / height: **5.16 / 2.33 / 2.80** (m)
Years of production: **1997 - 1998**

JCB

FASTRAC 2115

The 2115 used a slightly different 1000-6T HR1 engine producing 128 horsepower and 381lb-ft of torque at 1400rpm. The partially powered manual-shift gearbox now had 54 forward and 18 reverse gears in three overlapping ranges, although top speed remained the same at 37mph as it was limited by the hydraulic link between the steering wheel and the front wheels. Rear-axle- or four-wheel steering was available as an option on both the 2000 and 4000 series of JCBs, although it was only permitted below 12mph. An anti-lock braking system (ABS) was also an optional extra.

SPECIFICATION

Fuel: **Diesel**

Engine capacity: **6 litres (turbocharged)**

Cylinders: **6**

Power: **128hp**

Operating RPM: **1400**

Weight: **5930kg**

Top speed: **37mph**

Length / width / height: **5.08 / 2.25 / 2.88** (m)

Years of production: **1998 - 2004**

JCB
FASTRAC 2125

The JCB 2125 used a turbocharged six-litre Perkins 1000-6T HR2 engine developing 140 horsepower. Torque was back up to 434lb-ft but the gearing, top speed, PTO (540/1000rpm) and drive systems remained the same as on the 2115. The rear-lift capacity was also retained at 5251kg, with a hydraulic flow of 74 litres per minute.

SPECIFICATION

Fuel: **Diesel**
Engine capacity:
6 litres (turbocharged)
Cylinders: **6**
Power: **140hp**
Operating RPM:
1400 - 2300
Weight: **5931kg**
Top speed: **37mph**
Length / width / height:
5.08 / 2.25 / 2.88 (m)
Years of production:
1998 - 2004

JCB
FASTRAC 2135

The 2135 was supplied with a Perkins 1000-6TW HR5 engine developing 150 horsepower so torque was raised to 457lb-ft at 1500rpm. The four-wheel-steering system was also available on this model. It had five modes – two-wheel-steer, proportional steering, true-track, delay mode and crab steer – allowing the driver more manoeuvrability at low speeds. ABS was another factory option but an auxiliary hydraulic system was now fitted as standard.

SPECIFICATION

Fuel: **Diesel**

Engine capacity: **6 litres (turbocharged)**

Cylinders: **6**

Power: **150hp**

Operating RPM: **1400 - 2300**

Weight: **5931kg**

Top speed: **37mph**

Length / width / height: **5.08 / 2.25 / 2.88** (m)

Years of production: **1998 - 2004**

JCB
FASTRAC 2150

The four-wheel-steering option wasn't available on the Fastrac 2150 but it still used a 160 horsepower version of the 1000-6TW Perkins powerplant, although torque was up again to 489lb-ft at 1500rpm. The 54 forward and 18 reverse gears provided the same limited top speed as earlier models but the rear-lift capacity was increased to 6001kg. Although the general dimensions were the same as previous incarnations, the turning radius had increased from 13.5 metres in the 2115 to 14.2 metres in the 2150. This made the 2150 less manoeuvrable than its predecessors.

SPECIFICATION

Fuel: **Diesel**

Engine capacity: **6 litres (turbocharged)**

Cylinders: **6**

Power: **160hp**

Operating RPM: **1400 - 2200**

Weight: **6366kg**

Top speed: **37mph**

Length / width / height: **5.08 / 2.25 / 2.88** (m)

Years of production: **1998 - 2004**

JCB

FASTRAC 3155

Although production of the 3155 and the 2000 series ran concurrently, the new model was larger all round by a few centimetres and used a Perkins 1000-6 T4 engine developing 170 horsepower. Despite the 3155 being 400kg heavier, its torque was down on the 2150 (444lb-ft) and it reverted to the 36/12 gear arrangement so it wasn't as user-friendly. PTO was the same and the hydraulic system gave a rear-lift capacity of more than seven tonnes at 79 litres per minute so improvements were still being made.

SPECIFICATION

Fuel: **Diesel**

Engine capacity:
6 litres (turbocharged)

Cylinders: **6**

Power: **170hp**

Operating RPM:
1400 - 2400

Weight: **6766kg**

Top speed: **37mph**

Length / width / height:
5.15 / 2.30 / 3.15 (m)

Years of production:
1998 - 2002

JCB
FASTRAC 3185

The 3155 and 3185 were essentially upgraded versions of the earlier 155 and 185 respectively. The 3185 used a Cummins 6 BTA engine, however, which pushed out 190 horsepower. Fuel tank capacity was also increased to 350 litres (92 US gallons) and torque was up 20% to 553lb-ft at 1400rpm. The 3185 stuck with the 36 forward and 12 reverse gears, and it also retained the same hydraulic system, but the turning radius was back below 14 metres for extra manoeuvrability. From this model on, only Cummins engines were used.

SPECIFICATION

Fuel: **Diesel**
Engine capacity: **5.9 litres (turbocharged)**
Cylinders: **6**
Power: **190hp**
Operating RPM: **1400 - 2400**
Weight: **6766kg**
Top speed: **43mph**
Length / width / height: **5.15 / 2.49 / 3.15** (m)
Years of production: **1998 - 2002**

JCB
FASTRAC 2140

The Fastrac 2140 ushered in a new era for JCB. It was equipped with a turbocharged Cummins QSB30 water-cooled powerplant that delivered 440lb-ft of torque at 1500rpm. The 54/18 gearing configuration was reintroduced and there was an optional front PTO at 540/1000rpm. The hydraulics had been given an overhaul and employed a three-point hitch delivering a lift capacity of 6001kg. Hydropneumatic multi-link suspension – like that used in the Citroen range of cars – and hydrostatic steering provided comfort and safety, while rear-axle steering was also available as an option.

SPECIFICATION

Fuel: **Diesel**

Engine capacity: **6 litres (turbocharged)**

Cylinders: **6**

Power: **144hp**

Operating RPM: **1400 - 2200**

Weight: **6459kg**

Top speed: **37mph**

Length / width / height: **5.13 / 2.31 / 2.87** (m)

Years of production: **2005 - 2007**

JCB FASTRAC 8250

The enormous Fastrac 8250 was discontinued in 2011 but it remains the only model fitted with a CVT (continuously variable JCB V-Tronic) gearbox and an 8.3-litre Cummins QSC direct-injection common-rail turbo-diesel engine delivering 260 horsepower and 870lb-ft of torque. Fuel tank capacity was raised to 600 litres while the gearing allowed unlimited forward and reverse speeds and a creeper function. Electrohydraulic controls gave a rear-lift capacity of 10,000kg and a front capacity of 3500kg at 210 bar. The many accessories on this groundbreaking machine included: headland management, radar and performance monitor, high-flow hydraulics, electric mirrors, a winter pack, front and rear heated windows and mirrors, deluxe heated seat, climate control and cruise control. Such is its continuing popularity that good examples now trade for up to £70,000.

SPECIFICATION

Fuel: **Diesel**
Engine capacity: **8.3 litres (turbocharged with intercooler)**
Cylinders: **6**
Power: **260hp**
Operating RPM: **1300 - 2200**
Weight: **10,135kg**
Top speed: **42mph**
Length / width / height: **5.65 / 2.55 / 3.38** (m)
Years of production: **2005 - 2011**

JCB
FASTRAC 3170

The 3170 was another JCB to use the intercooled Cummins 6BTAA engine as it delivered excellent power and torque (580lb-ft) while also being economical. The tractor also had a redesigned bonnet so that the driver could check the service points more easily. The patented Smoothshift clutch reduced driver effort and was guaranteed to last for 6,000 hours, while an autoshift option allowed the engine to change automatically when under load.

The 3170 also came with four-link front suspension with a single control arm. The turning circle was improved by redesigning the upper arms to have greater clearance around the tyres. Anti-roll bars and self-levelling rear suspension improved handling, safety, traction and comfort, particularly when working on inclines. The two-seat cab had excellent visibility and employed noise-reduction insulation.

SPECIFICATION

Fuel: **Diesel**
Engine capacity: **5.9 litres (turbocharged)**
Cylinders: **6**
Power: **175hp**
Operating RPM: **1300 - 2200**
Weight: **7970kg**
Top speed: **41mph**
Length / width / height: **5.20 / 2.53 / 3.15** (m)
Years of production: **2005 – present**

JCB FASTRAC 3200 XTRA

The first versions of the 3200 Xtra were fitted with a 6x4 gearbox and a Cummins 6.7-litre powerplant but later models were some of the first to use the new 7.4-litre direct-injection diesel from AGCO Sisu. It gave a power hike to 223hp and increased torque to 708lb-ft. The 400-litre fuel tank and JCB P-Tronic transmission with 24 forward and nine reverse gears gave a top speed of 43mph and increased fuel economy over previous incarnations. Closed-centre load-sensing hydraulics gave lift capacities of: front 3500kg and rear 7000kg at 210 bar. Improved steering gave the enormous 3200 Xtra a turning radius of just 13.28 metres.

SPECIFICATION

Fuel: **Diesel**
Engine capacity: **7.4 litres (turbocharged with intercooler)**
Cylinders: **6**
Power: **223hp**
Operating RPM: **1500 - 2300**
Weight: **8610kg**
Top speed: **43mph**
Length / width / height: **5.20 / 2.52 / 3.21** (m)
Years of production: **2011 - present**

JCB

FASTRAC 3230 XTRA

The 3230 Xtra was essentially a beefed up version of the 3200. They were identical in size and used the same AGCO Sisu engine but the 3230 had 40 more horsepower and delivered a similar increase in torque, up 40lb-ft to 749lb-ft. The gearbox was the same so the top speeds were limited to 43mph. In fact the 3230 delivers the perfect balance of performance, comfort, efficiency and versatility. The power is available at the bottom of the rev range so it's quiet and fuel efficient, while the cabin is crammed with driver aids, safety features and the technology to make farming more efficient.

SPECIFICATION

Fuel: **Diesel**
Engine capacity:
7.4 litres (turbocharged with intercooler)
Cylinders: **6**
Power: **264hp**
Operating RPM:
1500 - 2300
Weight: **8610kg**
Top speed: **43mph**
Length / width / height:
5.20 / 2.52 / 3.21 (m)
Years of production:
2011 - present

A JCB Fastrac tractor busy cultivating

A JCB Fastrac makes light work of heavy snow

JCB tractors line up at the two-day Cereals event for arable farmers

A JCB Fastrac tractor with a square baler

A JCB Fastrac 8310 tractor with a manure spreader

Chapter 7

Leyland

Nuffield Tractors had been founded immediately after the Second World War by Lord Nuffield, who also owned Morris Motors Limited. The latter became part of the British Motor Corporation in 1951 and the first Leyland tractors appeared in 1969 when Leyland Motors bought BMC. In 1981, British Leyland was sold to Marshall & Sons who dropped the Leyland name for Marshall.

Leyland 344

By 1968 the tractor research team at the factory in Coventry had almost been phased out, but they would still contribute to the 344 and 384. Both were introduced to the farm machinery trade at Alexandra Palace, with the 344 being driven through a paper wall to make a memorable entrance. The press got their first look at the new machines at the Smithfield Show in December and the 344 went on sale to the public in January 1969 (the name came from the engine size and number of cylinders). It had an old BMC engine that had been developed in the 1940s, however, and an increase in crankshaft rotational speed to 2200rpm caused problems with imbalance and vibration, for which a balancer had to be fitted to reduce wear and tear.

SPECIFICATION

Fuel: **Diesel**
Engine capacity: **3.4 litres**
Cylinders: **4**
Power: **55hp**
Operating RPM: **1350 - 2200**
Weight: **2125kg**
Top speed: **18mph**
Years of production: **1969 - 1972**

Leyland 384

The 384 was launched at the same time as the slightly smaller 344. It had redesigned hydraulics to cope with heavier implements like ploughs and cultivating machinery. New controls allowed the driver to monitor plough draft and position but if they were incorrectly positioned the rear transmission tended to overheat and new hydraulic units had to be fitted to avoid costly lawsuits. Designer Mike Barnes later said that it wasn't until the company was bought by Marshall that an adequate hydraulic system was available. Both the 344 and 384 were developed and tested on endurance runs in Scotland to make sure they were ready for global distribution.

SPECIFICATION

Fuel: **Diesel**
Engine capacity: **3.8 litres**
Cylinders: **4**
Power: **70hp**
Operating RPM: **1400 - 2200**
Weight: **2585kg**
Top speed: **21mph**
Length / width / height: **3.40 / 1.89 / 2.48** (m)
Years of production: **1969 - 1972**

Leyland
154

The Leyland 154 began life as a Nuffield 4/25 (four cylinders/25 horsepower) but its predecessor hadn't lived up to expectations and a rebranding was necessary. The 154 was instead painted blue and given a new front grille, and it worked best with a Ransomes TS90 twin-furrow plough.

It was so reliable compared with its Japanese competitors that it was in production for 15 years but it was looking somewhat dated by 1979 and underwent a minor facelift. As it hadn't sold in the numbers predicted – the Japanese had eroded their chain of demand with inferior products – it was retired in favour of the 302 turbo-diesel.

SPECIFICATION

Fuel: **Diesel**

Engine capacity: **1.5 litres**

Cylinders: **4**

Power: **28hp**

Operating RPM: **1200 - 2500**

Weight: **1054kg**

Top speed: **11mph**

Length / width / height: **2.59 / 1.53 / 2.17** (m)

Years of production: **1970 - 1984**

Leyland
253

The 253 was also introduced at the annual Smithfield Show in 1971. It was billed as another all-new model that would appeal to farmers with medium-sized holdings, which were perhaps beyond the capability of the 344. The latest Perkins three-cylinder diesel wouldn't fit the original chassis so modifications had to be made to accommodate it. It was thought that making the engine and transmission part of the tractor body would reduce stress on the parts but several of the fixings were too weak and many tractors split at the engine-gearbox joint. The engine had also changed little in the last 20 years and was becoming outdated.

SPECIFICATION

Fuel: **Diesel**
Engine capacity: **2.5 litres**
Cylinders: **3**
Power: **35hp**
Operating RPM: **1200 - 2200**
Weight: **2086kg**
Top speed: **13mph**
Length / width / height: **3.20 / 1.62 / 2.30** (m)
Years of production: **1971 - 1972**

Leyland 245

As the 253 had been discovered to be a poor successor to the 154, it was rebadged and rebranded as the 245, which appeared the following year. It produced 50 horsepower from a similar powerplant as the 253, but the CVA DPA fuel pump would still be improved upon with the 255. As it was still essentially a small tractor, the frame provided only basic shelter and wasn't enclosed. In 1976, the 245 was fitted with a new internal hydraulic valve chest and a selector for the PTO.

SPECIFICATION

Fuel: **Diesel**
Engine capacity: **2.5 litres**
Cylinders: **3**
Power: **50hp**
Operating RPM: **1300 - 2250**
Weight: **2111kg**
Top speed: **16mph**
Length / width / height: **3.20 / 1.62 / 2.30** (m)
Years of production: **1972 - 1984**

Leyland 255/455

Leyland finally listened to the criticism of the 2.5-litre three-cylinder Perkins diesel and released two new tractors in 1972. The 255 and its four-wheel-drive cousin, the 455, were especially suited to work in larger fields and they became increasingly popular in Australia, where farmers liked the power hike because it meant they could use larger implements. These models were also well suited to driving front-mounted cultivators and rear harrows.

SPECIFICATION

Fuel: **Diesel**

Engine capacity: **3.6 litres**

Cylinders: **4**

Power: **60hp**

Operating RPM: **1400 - 2200**

Weight: **2439kg**

Top speed: **18mph**

Length / width / height: **3.34 / 1.62 / 2.35** (m)

Years of production: **1972 - 1976 / 1978 - 1981**

A Leyland 255

Leyland 270/470

The 270 and 470 (which had four-wheel-drive) replaced the outdated 384. This was another tractor that sold well in the Commonwealth as farmers in Australia, Canada and South Africa demanded more powerful machines that could harness larger tools. Traditional small-scale farming in the UK couldn't be overlooked, however, and the standard factory tractor also sold well at home.

A Leyland 270

SPECIFICATION

Fuel: **Diesel**

Engine capacity: **3.8 litres**

Cylinders: **4**

Power: **75hp**

Operating RPM: **1300 - 2200**

Weight: **2605kg / 3181kg**

Top speed: **20mph**

Length / width / height: **3.40 / 1.73 / 2.43** (m)

Years of production: **1972 - 1976 / 1973 - 1976**

Leyland
2100

Although most versions of the 2100 only put out 73 horsepower, demand from overseas forced Leyland to supply a variant with 100 horsepower. The tractor was similar to the 285 but it had the 6-98NT (normal tractor) engine rather than the 6-98DT (de-rated tractor) powerplant. The grip was exceptional considering it was only two-wheel-drive, but the clutch simply couldn't cope with the increase in power and torque and many had to be replaced. It was also something of an art form to operate the range and gear levers (which required both hands) while still trying to steer.

SPECIFICATION

Fuel: **Diesel**

Engine capacity: **5.6 litres**

Cylinders: **6**

Power: **100hp**

Operating RPM: **1200 - 2100**

Weight: **4041kg**

Top speed: **19mph**

Length / width / height: **3.81 / 2.17 / 2.28** (m)

Years of production: **1973 - 1979**

Leyland 262/462

Three years after the 98 series engine had been introduced, a more powerful version was installed but this meant the old tractors needed a facelift. Further changes were made to the transmission and hydraulics, which benefited from a new valve chest that was needed for working with larger implements. The tractor was shown to the public in November 1975 and released the following January. The four-wheel-drive version arrived in 1979.

A Leyland 262

SPECIFICATION

Fuel: **Diesel**
Engine capacity: **3.8 litres**
Cylinders: **4**
Power: **62hp**
Operating RPM: **1400 - 2200**
Weight: **2449kg / 3356kg**
Top speed: **15mph**
Length / width / height: **3.32 / 1.62 / 2.62** (m)
Years of production: **1976 - 1984 / 1979 - 1984**

Leyland 272/472

The increase in power and revised gearing to the rear pinion shaft led to the top speed of the 272 being limited to just under the legal limit for farm machinery of 25mph. A new cab added weight but also reduced the noise at the wheel from an illegal 91 decibels to an allowable 85. These new cabs were also significantly more robust than what had gone before. As the size and weight of tractors increased, drivers were more prone to injury in the event of a rollover and cabs were strengthened accordingly. This meant they could usually only be fitted by specialists and not lifted into place by a couple of farmhands. A synchromesh gearbox was introduced on most 272s from 1979.

SPECIFICATION

Fuel: **Diesel**
Engine capacity: **3.8 litres**
Cylinders: **4**
Power: **72hp**
Operating RPM: **1200 - 2000**
Weight: **3402kg / 3361kg**
Top speed: **23mph**
Length / width / height: **3.40 / 1.73 / 2.67** (m)
Years of production: **1976 - 1984**

A Leyland 472

Leyland
282/482

The 282 and its four-wheel-drive counterpart were delivered in configurations that allowed them to produce up to 100 horsepower, although the standard engine delivered 82. The introduction of the 4-98T (tractor) turbocharged engine accelerated development of these high-power machines and the export market was clearly a target for Leyland. The four-wheel-drive axles on the 482 were manufactured onsite by County Commercial Cars in Hampshire, while the front and rear wheels were the same size.

A Leyland 282

SPECIFICATION

Fuel: **Diesel**

Engine capacity: **3.8 litres (turbocharged)**

Cylinders: **4**

Power: **82hp**

Operating RPM: **1200 - 2200**

Weight: **3180kg / 3361kg**

Top speed: **20mph**

Length / width / height: **3.83 / 1.88 / 2.61** (m)

Years of production: **1979 - 1981**

Leyland 235

The Leyland 235 was another tractor that did well overseas. The export version had a 1.8-litre indirect-injection diesel producing 28 horsepower at 2500rpm. It had a CAV DPA hydraulic fuel pump, single dry plate clutch, nine forward and three reverse gears, hand-operated differential lock, adjustable free linkage draft sensors, lift capacity of 454kg, two-speed power takeoff (540/1000rpm), dry disc brakes and an adjustable swinging drawbar.

SPECIFICATION

Fuel: **Diesel**

Engine capacity: **1.8 litres**

Cylinders: **4**

Power: **35hp**

Operating RPM: **1200 - 2400**

Weight: **1229kg**

Top speed: **14mph**

Length / width / height: **2.60 / 1.55 / 1.45** (m)

Years of production: **1982 - 1984**

Leyland
602/604

The 602 and 604 tractors replaced the 262 and 462 in 1982. The 602 was usually fitted with the standard company cab, although some farmers sourced alternative Explorer cabs from Scandinavia. All the new tractors (602, 604, 702, 704, 802, 804) were given modern synchromesh gearboxes, oil-immersed brakes, depth-control hydraulics and independent multi-speed PTO, but production of the new range halved in the first few years of the 1980s as a global recession struck.

A Leyland 602

SPECIFICATION

Fuel: **Diesel**

Engine capacity: **3.8 litres**

Cylinders: **4**

Power: **62hp**

Operating RPM: **1200 - 2200**

Weight: **2903kg / 3050kg**

Top speed: **20mph**

Years of production: **1982 - 1984**

Leyland 702/704

The four-wheel-drive version of the 700 series was introduced a year before the 2WD model. They replaced the outdated 272 and 472 and used the same bigger engine as the 800 series. The new tractors were painted in a yellow livery and were immediately known as the Harvest Gold range. Most of the changes from previous incarnations were cosmetic but Leyland had invested heavily to fend off competition from JCB and they made small changes to the engine, gearbox and hydraulics.

SPECIFICATION

Fuel: **Diesel**
Engine capacity: **3.8 litres**
Cylinders: **4**
Power: **82hp**
Operating RPM: **1200 - 2200**
Weight: **2676kg / 2948kg**
Top speed: **21mph**
Years of production: **1983 - 1984 / 1982 - 1984**

A Leyland 704

Leyland
802/804

The 802 and 804 proved to be the last tractors the company produced before it was taken over by Marshall in the early 1980s. They could be fitted with a vented Sekura Explorer cab to keep the driver cool as both internal and external sources of heat couldn't be dissipated purely through the door. On hot days it was particularly uncomfortable but air-conditioning wouldn't be introduced until after the Leyland name had disappeared. The 4WD version was well suited to working with four-furrow reversible ploughs.

A Leyland 802

SPECIFICATION

Fuel: **Diesel**
Engine capacity: **3.8 litres**
Cylinders: **4**
Power: **82hp**
Operating RPM: **1200 - 2200**
Weight: **2676kg / 3039kg**
Top speed: **20mph**
Years of production: **1982 - 1984 / 1982 - 1985**

A Leyland Synchro with a flail mower attached

An old Leyland can still make light work of potato crops

A Leyland 245 proudly displayed

A retired Leyland after a hard working life

Chapter 8

Marshall

Marshall, Sons & Company was founded by William Marshall in 1842 when he bought an old engineering works in Gainsborough. He soon renamed it the Britannia Ironworks and began to develop steam engines. His sons, James and Henry, took over after his death in 1861.

Although the company initially produced traction engines and steamrollers, by 1900 they'd diversified into farm machinery (threshing machines in particular) and a brand of internal-combustion-engine-powered 'Colonial' tractor. At the outbreak of the First World War, the company was selling more than 300 units annually. The war slowed development but by 1928 they'd developed a tractor based on the German Lanz Bulldog, which Marshall called the 15/30.

Marshall COLONIAL

Although the output of the first Colonials was rated at 16+ horsepower, these figures are not comparable with the way power is measured now. When James Watt used the term in the late 18th century, he was comparing the work done by steam engines with that of horses, but he adapted his measurements to include piston engines, turbines and all kinds of farm machinery.

The Colonial was extremely popular in Canada and North America and soon earned a reputation as the 'Dreadnought of the Prairies'. Marshall seized on the phrase and was still marketing the steel-wheeled monster as such in 1920, at the same time warning potential buyers against succumbing to the lure of the new tractors.

SPECIFICATION

Fuel: **Petrol**

Cylinders: **4**

Power: **16-32hp**

Operating RPM: **400 - 600**

Weight: **9500kg**

Top speed: **4mph**

Years of production: **1908 - 1921**

Marshall 15/30 (E)

The 15/30 (E) was the first single-cylinder tractor released by Marshall. It had a bore of eight inches and a 10-inch stroke, which gave it around 10 horsepower, but it was underpowered and not as efficient as its rivals except in diesel form. It was supposed to be easy to start from cold but it was extremely difficult to crank the engine by hand while stuffing glowing ignition paper into the combustion chamber.

SPECIFICATION

Fuel: **Diesel**

Engine capacity: **8.2 litres**

Cylinders: **1**

Power: **10hp**

Operating RPM: **550**

Top speed: **4mph**

Years of production: **1930 - 1932**

Marshall 18/30

After just two years of production, the 15/30 was upgraded to become the 18/30. The one-cylinder diesel had the same bore and stroke but it could now rev to 700 so power increased a fraction. The transmission was also modified to increase torque – and therefore pulling power – at lower revolutions. The tractor was still far too heavy, however, so work began on the 12/20, although the company was also feeling the squeeze from the Great Depression and had to be rescued by Thomas Ward's scrap-iron business.

SPECIFICATION

Fuel: **Diesel**
Engine capacity: **8.2 litres**
Cylinders: **1**
Power: **13hp**
Operating RPM: **700**
Weight: **2650kg**
Top speed: **5mph**
Years of production: **1932 - 1936**

Marshall 12/20 (M)

The 12/20 was the next incarnation of the tried-and-tested Marshall. It had a completely new engine with a six-inch bore and nine-inch stroke, and it also had a new fuel-injection pump, cylinder head and transverse three-speed gearbox. The series had a facelift in 1938, which became the M, but production was severely retarded during the war as the company had to fulfil defence contracts. The M was produced for another two years and sold 1,000 units but it was then superseded by the Field-Marshall.

SPECIFICATION

Fuel: **Diesel**

Engine capacity: **5.1 litres**

Cylinders: **1**

Power: **32hp**

Operating RPM: **700**

Weight: **2246kg**

Top speed: **7mph**

Years of production: **1936 - 1938 / 1938 - 1946**

Marshall SERIES I

The Series I Field-Marshall came with two gearboxes delivering top speeds of either 6 or 9mph. The starting procedure still involved inserting a smouldering piece of paper infused with saltpetre into the cylinder head. The operator then had to turn the engine over by hand while a decompression valve removed air from the engine. After 3-6 revolutions, the flywheel had gained enough speed to turn the engine through compression and fire. It wasn't a satisfactory procedure and often gave the operator a complete workout before the engine finally turned over. Around 2,000 were sold for the relatively hefty price of £550.

SPECIFICATION

Fuel: **Diesel**

Engine capacity: **5.1 litres**

Cylinders: **1**

Power: **38hp**

Operating RPM: **750**

Top speed: **9mph**

Years of production: **1945 - 1947**

Marshall SERIES II

An option on the Series II addressed the starting issues by using a cartridge system to fire the engine. A shotgun cartridge was loaded into a breach on the engine intake while the smouldering paper was placed in the cylinder head. The operator then tapped the cartridge with a hammer (used as a firing pin) to blast a charge into the bore and force the piston into its stroke. Although it was far less strenuous than cranking the engine by hand, the cartridges left carbon deposits inside the engine, which then had to be cleaned. Despite the problems, more than 7,000 were sold.

SPECIFICATION

Fuel: **Diesel**

Engine capacity: **5.1 litres**

Cylinders: **1**

Power: **40hp**

Operating RPM: **750**

Top speed: **9mph**

Years of production: **1947 - 1949**

Marshall SERIES III

The starting issues wouldn't be resolved until the Series IIIA so the III was still hard work in every sense. When it was up and running, the tractor was mainly used for delivering threshing machines or other equipment from farm to farm as it didn't have the power for heavy fieldwork. It had a larger header tank, a centrally mounted PTO shaft, dual fuel filters, a stronger front axle and larger tyres to cope with the increased work. More than 3,200 were built but precious few survive today.

SPECIFICATION

Fuel: **Diesel**

Engine capacity: **5.1 litres**

Cylinders: **1**

Power: **40hp**

Operating RPM: **750**

Top speed: **9mph**

Years of production: **1949 - 1953**

Marshall SERIES IIIA

The Series IIIA was finally fitted with an electric starter, although it was still only an option. Marshall stuck with the distinctive two-stroke single-cylinder diesel coupled to an enormous flywheel that was still popular in Europe. (Competitors were rapidly switching to multi-cylinder engines as they delivered more power with a smoother and more linear torque curve.) The latest Marshall had a two-range hi-lo gearbox, which was extremely useful when the tractor was using its belt-pulley to drive other farm machinery. It also had a pressurised cooling system and could be fitted with a heavy-duty winch, although this bumped the price up to £845.

SPECIFICATION

Fuel: **Diesel**

Engine capacity: **5.1 litres**

Cylinders: **1**

Power: **42hp**

Operating RPM: **750**

Top speed: **10mph**

Years of production: **1953 - 1957**

Marshall
MP6

Marshall began development work on an MP4 in 1953 but only two prototypes were produced. They were powered by a 65-horsepower Meadows 4DC330 diesel unit with six forward and two reverse speeds. The MP4 didn't make it into production, however, as the Leyland UE350 diesel-powered MP6 immediately rendered it obsolete. The tractor was built primarily for the sugar cane industry overseas. Although it was advertised as a heavy-duty machine, transmission failures were common and it didn't have a hydraulic lift. As a result, only 197 were sold by 1961 and the line was discontinued.

SPECIFICATION

Fuel: **Diesel**

Engine capacity: **5.8 litres**

Cylinders: **6**

Power: **70hp**

Operating RPM: **1700**

Top speed: **13mph**

Years of production: **1954 - 1961**

Chapter 9

Muir-Hill

Muir-Hill was an engineering company based near Old Trafford in Manchester. From the early 1920s, the company produced locomotives and high-power four-wheel drive tractors. The locomotive side of the operation wound down in the 1930s so they could concentrate on the manufacture of construction equipment and hydraulic loaders like forklifts.

After the war, Muir-Hill began to focus solely on the tractor arm of the business, with the first four-wheel-drive variant coming to market in 1969. The 4WD system ran from the sandwich plate between the gearbox and the rear axle so the ride height had to be elevated. This gave the series great ground clearance and also improved traction through its four equal-sized wheels. The tractors were most commonly used for heavy draft applications, although a downturn in company fortunes meant that production stopped in the mid 1980s. Muir-Hill is now owned by Lloyd Loaders.

Muir-Hill

101

David Brown introduced tractors with new loading shovels in the early 1960s. As they boasted lifting capacities up to three tonnes, the powerful Muir-Hill tractor range with their 108-horsepower six-cylinder Perkins or Ford diesels were ideal for running these heavy agricultural and industrial applications. The 101 was introduced to the public in 1966 but it didn't go on general sale for another 18 months. Two variations were released towards the end of the decade, and they were usually seen pulling four-furrow Lemken Reversible Ploughs or Dowdswell five-furrow semi-mounted ploughs, even though the latter could cause instability when turning the plough over.

SPECIFICATION

Fuel: **Diesel**

Engine capacity: **5.9 litres**

Cylinders: **6**

Power: **108hp**

Operating RPM: **1500 - 2500**

Weight: **4371kg**

Top speed: **18mph**

Length / width / height: **3.93 / 2.19 / 2.03** (m)

Years of production: **1968 - 1972**

Muir-Hill
110

The Muir-Hill 110 had a slightly smaller and less powerful Perkins A6/354 diesel driving all four wheels. It was essentially the same machine as the 101 but was instead mainly sold on the export market. The 101 and 110 were capable of hauling 13- and 20-tonne trailer dumpers but a new tractor was needed when the trailers topped 25 tonnes. From this customer need, the 161 evolved, which became the most powerful tractor in Britain.

SPECIFICATION

Fuel: **Diesel**
Engine capacity: **5.8 litres**
Cylinders: **6**
Power: **84hp**
Operating RPM: **1500 - 2500**
Weight: **4116kg**
Top speed: **15mph**
Length / width / height: **3.93 / 2.19 / 2.08** (m)
Years of production: **1969 - 1972**

Muir-Hill
111

The Muir-Hill 111 was the most powerful tractor yet built by the company. It was designed to replace the less sturdy 110 so the Perkins A6/354 had to be adapted for harder agricultural and industrial use. The 111 proved to be the company's workhorse and it remained in production for more than a decade.

SPECIFICATION

Fuel: **Diesel**
Engine capacity: **5.8 litres**
Cylinders: **6**
Power: **110hp**
Operating RPM: **1400 - 2500**
Weight: **4853kg**
Top speed: **19mph**
Length / width / height: **3.96 / 2.08 / 2.03** (m)
Years of production: **1972 - 1983**

Muir-Hill

121

The MH 121 was designed as a replacement for the 101, although it was somewhat strangely called the Series II (no Series I was ever made). The 121 featured a walk-through flat floor and a fully glazed cab to keep out the elements so it was extremely popular with consumers despite its lack of power. More innovations in terms of usability and comfort were needed before the brand would be seen as a serious rival to JCB and Leyland, however. The Series III addressed these issues with a larger cab, soundproofing, air conditioning and a radio.

SPECIFICATION

Fuel: **Diesel**

Engine capacity: **6.2 litres**

Cylinders: **6**

Power: **120hp**

Operating RPM: **1600 - 2500**

Weight: **4853kg**

Top speed: **19mph**

Length / width / height: **3.96 / 2.08 / 2.03** (m)

Years of production: **1972 - 1983**

Muir-Hill
141

The Muir-Hill brand was under threat by the end of the 1970s but it was saved by the intervention of Babcock Construction Limited. The first tractor from the Series III range was the 141, which had a choice of engines from the Ford 2704ET to the Perkins T6/354/4. It benefited from more creature comfort than were installed in the 121 and soon earned a reputation as a reliable and powerful workhorse. Many examples can still be found working the land today.

SPECIFICATION

Fuel: **Diesel**

Engine capacity: **5.9 litres (turbocharged)**

Cylinders: **6**

Power: **145hp**

Operating RPM: **1500 - 2500**

Weight: **5896kg**

Top speed: **19mph**

Years of production: **1978 - 1983**

Muir-Hill
161

The 161 continued the evolution of the Muir-Hill brand by incorporating a big Perkins V8 (540) diesel pumping out 185 horsepower, which made it the most powerful tractor in the country. While the smaller tractors in the series were only rated to tow up to 20 tonnes, the 161 was capable of towing 25 tonnes at more than 20mph. It therefore made perfect sense for transporting large industrial and agricultural loads and could often be seen being driven around building sites, road works and amenity tips.

SPECIFICATION

Fuel: **Diesel**
Engine capacity: **8.4 litres**
Cylinders: **8**
Power: **185hp**
Operating RPM: **1500 - 2500**
Weight: **5443kg**
Top speed: **21mph**
Length / width / height: **4.19 / 2.47 / 2.89** (m)
Years of production: **1969 - 1975**

Muir-Hill
171

The MH 171 was much the same as the 161, although the engine wasn't quite as powerful and the cab had been modified to make the controls easier to use. It's not that easy to tell them apart from distance, but the 171 has twin exhaust stacks at the front of the bonnet. They were used everywhere from sugar cane plantations in Guyana to the prairies of North America, from the Australian outback to launching boats in the Caribbean.

SPECIFICATION

Fuel: **Diesel**

Engine capacity: **8.8 litres**

Cylinders: **8**

Power: **177hp**

Operating RPM: **1600 - 2500**

Weight: **6350kg**

Top speed: **20mph**

Length / width / height: **4.73 / 2.45 / 3.13** (m)

Years of production: **1975 - 1984**

Chapter 10

Nuffield

Morris Motors developed the Nuffield tractor after the Second World War when there was high demand for farm machinery and a need to export equipment to replenish the country's coffers. The prototype Nuffield 'Universal' was running by the end of 1945 but a shortage of raw materials meant the project had to be shelved for two years. By the end of 1948, the company was ready to launch the tractor in 'poppy orange' livery. It was powered by a four-cylinder Morris commercial truck engine that had been modified to run on TVO. The Nuffield brand lasted until 1969 when it was bought by Leyland.

Nuffield M3

The M3 had a single front wheel carried in an inverted U-shaped fork and, with the adjustable rear hubs clamped to the axle shaft, it was clearly designed for rowcrop work. A clever front axle meant that the single wheel could be swapped for two wheels in less than half an hour so it was a popular tractor for both farming and industrial uses. The engine had been used in Morris's military trucks so it had a main carburettor for tractor vaporising oil, although it also had an auxiliary 'starter' carburettor when burning petrol. The powerplant was mated to a five-speed gearbox and there was an optional hydraulic lift. A petrol version of the M3 (PM3) was launched in 1950, with the diesel variant (DM3) following later in the year.

SPECIFICATION

Fuel: **Tractor Vaporising Oil**
Engine capacity: **3.4 litres**
Cylinders: **4**
Power: **38hp**
Operating RPM: **1000**
Weight: **2450kg**
Top speed: **16mph**
Years of production: **1948 - 1956**

Nuffield M4

The Universal M4 was the four-wheeled equivalent of the M3. It also had a durable cast-iron frame plus six more sub-assemblies: ETA designated engine, clutch system, steering box, transmission tunnel and two reduction units. All of these could be removed individually so the tractor was easy to maintain and cheap to run. Optional extras included the hydraulic lift assembly, PTO and braking systems.

The diesel DM4 arrived in 1950. It had a few extra horsepower but wasn't as popular and was discontinued in 1954. The petrol PM4 had 48 horsepower and production lasted until 1957. Midway through its run, its old spark-ignition power units were replaced with ETC (for the TVO) and ETD (for the petrol) engines delivering a higher compression ratio and even more power.

SPECIFICATION

Fuel: **Tractor Vaporising Oil**
Engine capacity: **3.4 litres**
Cylinders: **4**
Power: **38**
Operating RPM: **1000**
Weight: **2570kg**
Top speed: **16mph**
Years of production: **1948 - 1956**

Nuffield
UNIVERSAL III

The Universal III was designed to make a dent in the sales of the Ferguson 35 and the Fordson Dexta, but it was much heavier and was slightly underpowered by its BMC OEC/2(B) diesel. Nuffield tried to increase power with a four-bearing crankshaft and a Simms injection pump, and they also went with a small three-cylinder engine to reduce weight. Despite its shortcomings, Nuffield had redesigned the driving position with comfort and safety as priorities, and the steering was so light that it proved a popular tractor. The four-cylinder Universal IV followed quickly on its heels and the 56-horsepower engine immediately rendered the III obsolete.

SPECIFICATION

Fuel: **Diesel**

Engine capacity: **2.6 litres**

Cylinders: **3**

Power: **37hp**

Operating RPM: **1000 - 2000**

Top speed: **15mph**

Years of production: **1957 - 1961**

Nuffield
3/42

The 3/42 was manufactured at Nuffield's new premises in Bathgate in Scotland. Power was up a little on the standard Universal III but a hydraulic depth-control system with top-link sensing had been added. This hydraulic power unit could be controlled separately from the PTO. As the controls were well laid out, visibility was good all round, the clutch action was smooth, the driving position was uncluttered and free from moving parts, and the rubber-mounted seat gave excellent support on all lateral gradients, the 3/42 was a popular addition to the Nuffield fleet.

SPECIFICATION

Fuel: **Diesel**
Engine capacity: **2.8 litres**
Cylinders: **3**
Power: **39hp**
Operating RPM: **1000 - 1600**
Weight: **1996kg**
Top speed: **16mph**
Length / width / height: **2.95 / 1.62 / 2.09** (m)
Years of production: **1961 - 1963**

Nuffield 4/60

The 4/60 was simply the four-cylinder version of the 3/42, although the horsepower claimed never quite reached the 60 suggested by its name. The engine was a new OEE version of the standard BMC diesel with an increased bore (100mm as opposed to 95mm), while later models had the double clutch OEE/3 power unit. A Simms Minimec fuel-injection pump was also added to try to reach the magical 60hp but few farmers reported much difference in power. The last one off the Ward End production line in Birmingham was exported to Australia in 1963 (chassis number T41481).

SPECIFICATION

Fuel: **Diesel**

Engine capacity: **3.8 litres**

Cylinders: **4**

Power: **55hp**

Operating RPM: **1000 - 2000**

Weight: **2580kg**

Top speed: **17mph**

Length / width / height: **2.95 / 1.62 / 2.09** (m)

Years of production: **1961 - 1963**

Nuffield 10/42

The 10/42 and 10/60 models were built to replace the 3/42 and the 4/60. Power at last reflected the name and the 10/42 also came with a new dual-flow hydraulic system that finally addressed the problems with earlier models. The 10-speed gearbox was another welcome upgrade as it allowed farmers to maintain a variety of speeds depending on gradient, conditions and crop.

The 10/42 may have been smaller and less powerful than its big brother but a Minimec pump was an optional extra. The models also featured the latest self-energising disc brakes, a redesigned drawbar and a cleaner instrument panel.

SPECIFICATION

Fuel: **Diesel**
Engine capacity: **2.8 litres**
Cylinders: **3**
Power: **42hp**
Operating RPM: **1000 - 2000**
Weight: **2417kg**
Top speed: **15mph**
Years of production: **1964 - 1966**

Nuffield 10/60

The four-cylinder 10/60 had a modified gearbox and hydraulic system but power was unchanged from the 4/60. To get 10 gears from the box, Nuffield used the old five-speed unit but added a lay shaft and splitter gear to the main drive shaft, which doubled the gears available. A sliding sleeve housed the splitter gear and was splined to the main shaft. A hi-lo lever next to the primary gear lever allowed the driver to alternate between the two and monitor speed more accurately.

SPECIFICATION

Fuel: **Diesel**
Engine capacity: **3.8 litres**
Cylinders: **4**
Power: **60hp**
Operating RPM: **1000 - 2000**
Weight: **2660kg**
Top speed: **17mph**
Years of production: **1964 - 1967**

Nuffield 3/45

The Nuffield 3/45 had a water-cooled three-cylinder BMC engine developing 45 horsepower (a slightly different version was available for the US market). It retained the popular 10-speed gearbox of its predecessor and was extremely manoeuvrable with its ground clearance of 36cm and turning radius of just 3.4 metres.

SPECIFICATION

Fuel: **Diesel**

Engine capacity: **2.8 litres**

Cylinders: **3**

Power: **45hp**

Operating RPM: **1000 - 1600**

Weight: **2023kg**

Top speed: **16mph**

Length / width / height: **2.95 / 1.62 / 2.09** (m)

Years of production: **1967 - 1969**

Nuffield

4/25

The 4/25 may have been released alongside the 3/45 but it was a poor second to its stable mate. Petrol wasn't as popular as diesel and the small 1.8-litre engine only pushed out 25 horsepower. As such, it simply didn't have enough grunt for larger agricultural or industrial applications and was only suitable for mowing and towing smaller loads. Independent PTO at 540 and 1000rpm was also standard by now so this 2WD model with nine forward and two reverse gears offered little that wasn't available elsewhere.

SPECIFICATION

Fuel: **Petrol**

Engine capacity: **1.8 litres**

Cylinders: **4**

Power: **25hp**

Operating RPM: **1000 - 2500**

Weight: **1056kg**

Top speed: **15mph**

Years of production: **1967 - 1969**

Nuffield 4/65

The 4/65 was only produced at the Bathgate factory in Scotland for two years but it was a worthy replacement for the 10/60. It had a twin-range 10-speed gearbox, new rear mudguards, a restyled radiator grille with built-in lights, a bigger radiator and a larger fuel tank in the nose. It also had a tandem hydraulic pump with the controls altered to a quadrant design. Bray Engineers offered to convert it to 4WD and add power steering, although the first option didn't prove popular (130 units sold in three years) and the second option needed an expensive pump mounted on the engine to run properly. Some restored examples are worth £5,000 today.

SPECIFICATION

Fuel: **Diesel**
Engine capacity: **3.8 litres**
Cylinders: **4**
Power: **65hp**
Operating RPM: **1200 - 2200**
Weight: **2331kg**
Top speed: **19mph**
Length / width / height: **3.21 / 1.62 / 2.20** (m)
Years of production: **1967 - 1969**

A Nuffield tractor at Cophill Farm vintage rally

A Simms fuel injection pump on a Nuffield diesel tractor

A Nuffield 460 tractor taking part in a potato picking demonstration

An old Nuffield tractor still going strong after 50 years

This Nuffield M4 Universal tractor was converted by Perkins in the early 1950s to use an L4 diesel engine

Chapter 11

Saunderson

Herbert Percy Saunderson was born in Bedfordshire in 1869. He briefly worked for Massey-Harris in Canada before returning to Britain and founding a tractor and implement company in Bedford in 1890. Eight years later he released a self-moving vehicle but it was unreliable and few units were sold. By 1904 he'd built his first tractor and the company was soon the largest tractor-building operation in the world. The Model G was particularly popular and the brand remained the public's first choice until the arrival of Fordson, which negatively impacted on sales at home and abroad. In 1924, Saunderson sold the business to Crossley Motors but the union only lasted until the mid 1930s before another merger resulted in the Bedford Plough & Engineering Company. The latest incarnation survived until 1970.

Saunderson
MODEL A

The Model A was a three-wheeled tractor that appeared two years after Saunderson's first farm machines. Production only lasted around four years as the four-wheel variants surpassed the Model A for practicality, safety and power. The first versions were slow and cumbersome and were so complicated that they didn't threaten the dominance of the traditional horse on many farms.

The drive gears were protected by pressed steel and a spring was provided for driver comfort. A tubular radiator with a belt-driven fan was used for cooling. The tractor had a passenger seat opposite the driver and a removable load-carrying platform at the back. The Type A later became known as the Model U with minor modifications. A restored Model A from 1906 is on display in Swan Hill, Victoria, Australia, one of precious few to survive today.

SPECIFICATION

Fuel: **Petrol**
Engine capacity: **11.1 litres**
Cylinders: **4**
Power: **50hp**
Operating RPM: **600 - 800**
Top speed: **7mph**
Years of production: **1906 - 1910**

Saunderson MODEL L

The Model L and its big brother (the Model F) were introduced by Saunderson & Gifkins when the world was at war. The agricultural industry in the UK was in crisis but the new models were useful in the fields and could also be fitted with a removable truck body for transporting loads by road. This proved invaluable when delivering supplies, ammunition and men to the coast for their postings to northern France.

SPECIFICATION

Fuel: **Petrol**

Engine capacity: **11 litres**

Cylinders: **1**

Power: **7hp**

Operating RPM: **750**

Weight: **1100kg**

Top speed: **7mph**

Years of production: **1910 - 1914**

Saunderson MODEL F

The Model L was a 'little universal tractor' with a single-cylinder seven-horsepower engine but it was soon replaced with the Model F. The F had a two-cylinder water-cooled powerplant producing twice the output. It boasted a special single-furrow plough, and a threshing machine could be run from the belt pulley. The driver originally sat at the front of the tractor but this was later moved back and to the right-hand side.

SPECIFICATION

Fuel: **Petrol**

Engine capacity: **6 litres**

Cylinders: **2**

Power: **14hp**

Operating RPM: **600 - 800**

Weight: **1200kg**

Top speed: **5mph**

Years of production: **1916 - 1920**

Saunderson MODEL G

The Model G was a cumbersome brute with two enormous steel wheels at the rear, a seat open to the elements, a large and inefficient engine and two tiny steel wheels at the front. Although the wheels had raised tread, they were inadequate in soft or boggy ground. Despite its shortcomings, this was the most successful Saunderson tractor. A restored Model G can be found in the Grampian Transport Museum near Aberdeen in Scotland. Another has been preserved at the Rutland County Museum in England.

SPECIFICATION

Fuel: **Petrol**

Engine capacity: **6.2 litres**

Cylinders: **2**

Power: **25hp**

Operating RPM: **750**

Top speed: **5mph**

Years of production: **1916 - 1924**

Saunderson LIGHT

The Saunderson Light Tractor was introduced when the company was already in financial difficulty. The war had been a drain on men and resources, and the agricultural sector was still in decline. A super lightweight version with two cylinders and a 20-horsepower engine combined with a gearbox was released in 1923, but it cost £195 and few were sold.

When the Crossley brothers took over the business in 1924, the tractors were all rebranded, but the company then began to focus on petrol-powered locomotives. The workshop in Bedford continued to produce Saunderson tractors, engines and agricultural machinery for another decade or so before it became Bedford Plough & Engineering.

SPECIFICATION

Fuel: **Petrol/Paraffin**
Engine capacity: **6 litres**
Cylinders: **2**
Power: **20hp**
Operating RPM: **750**
Top speed: **5mph**
Years of production: **1922 - 1924**

Saunderson
MODEL V

The Model V was a huge four-cylinder monster that was Saunderson's flagship model. It developed 45-50bhp from its four-cylinder 6x8 engine. The Model V was extremely popular with the overseas market, particularly the Commonwealth, although photos also show them working in Russia and West Africa. Saunderson & Gifkins later signed an agreement to ship them out to central and eastern Europe.

SPECIFICATION

Fuel: **Petrol**

Engine capacity: **10 litres**

Cylinders: **4**

Power: **50hp**

Operating RPM: **600 - 800**

Weight: **1800kg**

Top speed: **8mph**

Years of production: **1912 - 1914**

The pictures in this book were provided courtesy of the following:

WIKIMEDIA COMMONS, FLICKR CREATIVE COMMONS

Design & Artwork: ALEX YOUNG

Published by: G2 ENTERTAINMENT LIMITED

Publisher: EDWARD ADAMS & JULES GAMMOND

Printed in Dunstable, United Kingdom

91222432R00074